How Mobile Is the Footloose Industry?
The Case of the Notebook PC Industry in China

Policy Studies
an East-West Center series

Series Editors
Edward Aspinall and Dieter Ernst

Description
Policy Studies presents scholarly analysis of key contemporary domestic and international political, economic, and strategic issues affecting Asia in a policy relevant manner. Written for the policy community, academics, journalists, and the informed public, the peer-reviewed publications in this series provide new policy insights and perspectives based on extensive fieldwork and rigorous scholarship.

THOMSON REUTERS
BOOK CITATION INDEX
INDEXED

The East-West Center is pleased to announce that the Policy Studies series has been accepted for indexing in Web of Science Book Citation Index. The Web of Science is the largest and most comprehensive citation index available.

Notes to Contributors
Submissions may take the form of a proposal or complete manuscript. For more information on the Policy Studies series, please contact the Series Editors.

Editors, Policy Studies
East-West Center
1601 East-West Road
Honolulu, Hawai'i 96848-1601
Tel: 808.944.7197
Publications@EastWestCenter.org
EastWestCenter.org/PolicyStudies

Policy
Studies | 67

How Mobile Is the Footloose Industry?

The Case of the Notebook PC Industry in China

Tain-Jy Chen and Ying-Hua Ku

How Mobile Is the Footloose Industry?
The Case of the Notebook PC Industry in China
Tain-Jy Chen and Ying-Hua Ku

ISSN 1547-1349 (print) and 1547-1330 (electronic)
ISBN 978-0-86638-241-0 (print) and 978-0-86638-242-7 (electronic)

The views expressed are those of the author(s) and not necessarily those of the Center.

Hard copies of all titles, and free electronic copies of most titles, are available from:

Publication Sales Office
East-West Center
1601 East-West Road
Honolulu, Hawai'i 96848-1601
Tel: 808.944.7145
Fax: 808.944.7376
EWCBooks@EastWestCenter.org
EastWestCenter.org/PolicyStudies

In Asia, hard copies of all titles, and electronic copies of select South-east Asia titles, co-published in Singapore, are available from:

Institute of Southeast Asian Studies
30 Heng Mui Keng Terrace
Pasir Panjang Road, Singapore 119614
publish@iseas.edu.sg
bookshop.iseas.edu.sg

Contents

List of Acronyms

CPU   central processing unit

GDP   gross domestic product

GPN   global production network

HP   Hewlett-Packard

MNC   multinational corporation

PC   personal computer

R&D   research and development

VAT   value added tax

VMI   vendor managed inventory

Executive Summary

In this issue, a case study of the notebook personal computer (PC) industry in China is used to reexamine the concept of the "footloose industry." As labor costs rose dramatically in the coastal areas of China where production has been concentrated, contract manufacturers attempted to move their factories to the inner provinces of China, in response to incentives from the host regions' governments. However, they encountered immense difficulties in recruiting labor and managing supply chains in the new locations, despite the generous support offered by the local governments. The experience highlights the important role of labor institutions in which global production networks (GPNs) are embedded. Labor institutions and GPNs are interdependent and coevolving. Unique Chinese labor institutions have reshaped the structure of GPNs of the notebook PC industry, which was brought to China by multinational corporations (MNCs) in the 1990s. Because the labor institutions of the coastal areas cannot be replicated in the inner provinces, it is difficult to relocate the production there. This makes a footloose industry like the notebook PC industry not so loose anymore.

Unique Chinese labor institutions have reshaped the notebook PC industry

The traditional argument that labor-intensive industries operated by MNCs are footloose is based on the assumption that labor is homogeneous and location-bound, while capital is perfectly mobile. Therefore, capital moves freely to exploit location-bound labor. This assumption is untrue, as labor is mostly heterogeneous and potentially mobile. This case study shows that even when labor is homogeneous, which may be close to reality in the case of unskilled workers, the sheer size of the labor pool still makes a difference. Moreover, the mobility of labor, which can be controlled by migration policy and related labor market regulations, also matters. The quantity of labor affects the scale of production, which, in turn, affects the organization of production along the value chains and the power structure of GPNs. The mobility of labor affects the internal organization of work within the firm. When both the inter- and intra-firm organization of production are conditional on local labor institutions, it is not easy for the production activities to be relocated.

Labor institutions found only in the coastal areas of China have significantly reduced the number of players in the GPNs, making the notebook PC industry more concentrated than before. Both the number of contract manufacturers and their first-tier suppliers have significantly decreased, and the power structure of the GPNs has shifted in favor of the contract manufacturers. Along with the shrinking number of players, production has become more vertically integrated, and the alliances between the contract manufacturers and their suppliers have been strengthened. As a result, the contract manufacturers have had more power in making location decisions, and they respond to local government incentives. However, as GPNs are deeply embedded in local institutions, relocation entails a search for the right local government that is willing and able to create the right institutions to accommodate the new GPN structures. Because a perfect replication of local institutions is difficult, a wholesale re-location of GPNs from one place to another is also implausible. A good coupling of GPNs with local institutions is a precondition for global factories to function. This presents a big challenge to the concept of the footloose industry, which implies that the industry can move without friction.

Once rising labor costs affected the coastal areas of China, there is no doubt that the region lost its location advantage in hosting the global factories. The institutions underscoring this location advantage have become obsolete, and they have to be restructured in order to sustain the growth momentum of the region. At the same time, regions that are interested in hosting the global factories currently located in the coastal areas need to construct a set of effective institutions. These institutions cannot be a replica of coastal China's institutions, but must be new and idiosyncratic, with the potential to reshape the nature of global competition. Therefore, both the coastal and inland areas of China are facing the challenge of implementing institutional reforms to retain or attract the so-called footloose industries. Indeed, competition among local institutions has been an important part of the regional rivalry in China since the economic reforms began in 1978.

Regional competition led to innovations in institutions, which have chartered the course of global competition

Regional competition often led to innovations in institutions, which have chartered the course of global competition in the last 30 years. The game seems to have been reset.

How Mobile Is the Footloose Industry?
The Case of the Notebook PC Industry in China

Introduction

It is well established in the literature that labor-intensive industries established by multinational corporations (MNCs) in developing countries are sensitive to local wages. MNCs come when local wages are low, and their presence tends to drive up local wages. When local wages rise to a certain level, they move out to seek a new location that offers lower wages. This type of industry is therefore called a "footloose" industry (Flamm 1984), as it provides only transitory benefits to the local economy. By repeatedly moving around the globe to exploit low-cost labor, footloose industries impose ceilings on local wages, which are mitigated only when developing countries find alternative sources to enhance their labor productivity. This phenomenon is sometimes referred to as a "race to the bottom" in the literature (Klevorick 1996; Mehmet and Tavakoli 2003; Rudra 2008).

Moving around the globe to exploit low-cost labor, footloose industries impose a 'race to the bottom'

However, recent developments in global production networks (GPNs) may have fundamentally changed the nature of footloose industries. In short, expanding globalization has increased the market concentration of suppliers in the GPNs, allowing them to command larger market shares. As a result, the suppliers or their contract manufacturers have become more vertically integrated in the production processes in order to exploit scale economies (Appelbaum 2008). There is also a tendency for the smaller number of manufacturers in the same industry to cluster in a specific region, together with their components and parts providers. Clustering reduces the costs of coordination, which is essential to the system of massive and flexible production that characterizes today's global supply chains (Baldwin and Venables 2010). "Massive" production is needed because the entire world demand is supplied by a few clusters, and sometimes only one cluster, in a particular region. "Flexible" production is needed because consumer preferences vary over time and responsiveness to changing consumer demand is essential to global competition. Underscoring these clusters are enormous numbers of workers, both skilled and unskilled, who can be hired and fired with ease. This makes the traditional footloose industry no longer footloose, as the size of the local labor supply and the labor institutions that determine the flexibility of production become binding constraints on the mobility of global factories. When the local wages increase, the industry may be trapped in the existing location, unable to find a new home.

This paper presents a case study of the notebook personal computer (PC) industry, whose production has been concentrated in the eastern coastal areas of China in recent years. As labor costs rose dramatically in the region, contract manufacturers attempted to move to the inner provinces of China, following the initiatives of the host regions' governments. However, they encountered immense difficulties in recruiting labor and managing supply chains in the new locations, despite the generous support offered by local governments. The experience highlights the important role of local labor markets in global production networks. Simply put, the Chinese labor market, which is unique in terms of size and mobility, has reshaped the structure of GPNs, making the footloose industries far less itinerant than is their norm. To the extent that labor size and mobility are important to the location decision of global production, the case of the notebook PC industry presented in this paper exemplifies the new realities facing the traditional footloose industries.

Global Production Networks

A global production network (GPN) is based on the concept that production in a globalized world is accomplished sequentially by a network of firms specializing in different stages of production and located in different regions. A chain of production spanning R&D, design, manufacturing, marketing, and after-sales service is organized by a group of firms that are geographically scattered, but closely coordinated. Vertical fragmentation allows the industry in question to exploit location-specific resources in different regions. The location of production is determined by a complex set of factors, including the power structure of the GPNs, firm capacity, state policy, and local institutions (Coe, Dicken, and Hess 2008). Scholars who study GPNs tend to highlight the aspects of their particular interests. The existing literature can, by and large, be separated into three schools: sociological, geographic, and economics schools. The sociological school focuses on power distribution and firm capacity, as they are concerned about the distribution of benefits coming out of global production (Gereffi 1994, 1995; Gereffi, Humphrey, and Sturgeon 2005). The geographic school focuses on the roles of state and regional capacity, as they are concerned with the spatial distribution of global production (Coe, Dicken, and Hess 2008; Hess and Yeung 2006). The economics school, meanwhile, is concerned with the transaction costs in global production, which are affected by local institutions and geographical distance between producers (Coase 1937; Williamson 1975, 2008). It goes without saying that the three schools are closely connected and complementary to one another. This paper argues that state policy and regional capacity can also affect the transaction costs in the region and the power structure of GPNs. In particular, it highlights the role of local government as a power broker of the GPNs, tilting the benefits toward a certain group of firms that are strategically coupled with the locality. When GPNs are embedded in specific local institutions created by local governments, it becomes difficult to change the spatial distribution of global production, especially when these institutions are not easily replicated elsewhere. Without these location-specific institutions, transaction costs become too high to bear. In essence, GPNs are not static, but are dynamic in nature. They coevolve with local institutions, and are therefore path-dependent.

One specific local institution that this paper analyzes is the labor market. The labor market has been acknowledged in the GPN literature as being a critical local institution that affects the spatial distribution of global production (Coe, Dicken, and Hess 2008; Rainnie, Herod, and McGrath-Champ 2010). Labor is an input of production, as well as a consumer in the global market. Labor is spatially differentiated in terms of quantity and quality. Labor is an embodiment of local institutions, including cultural and regulatory institutions. The nature of the local labor market is an essential determinant of the spatial distribution of global production (Riisgaard and Hammer 2011), especially for the labor-intensive production that is the focus of this paper.

> *The local labor market is an essential determinant of the spatial distribution of global production*

The argument that labor-intensive industries operated by MNCs are footloose is based on the assumption that labor is mostly homogeneous and location-bound. By contrast, capital is perfectly mobile. This assertion is, of course, untrue, as labor is idiosyncratic and potentially mobile. The following case study will show that even when labor is homogeneous, which may be largely true in the case of unskilled workers, the sheer size of the labor pool still makes a difference. Moreover, the mobility of labor, which can be controlled by migration policy and related labor market regulations, also matters. The quantity of labor affects the scale of production, which, in turn, affects the organization of production along the value chains and the structure of the GPNs. On the other hand, the mobility of labor affects the internal organization of work within the firm. When the organization of production both among firms and within each firm are conditional on the local labor market, it is difficult for production activities to be relocated.

The notebook PC industry in China will be used to illustrate these points. Traditionally, the GPNs of the PC industry have been dominated by the brand marketers, which are considered to be the flagship firms (Ernst and Kim 2002). However, China's immense labor supply has changed this assertion. Aided by the scale economies derived from a large labor force that is unmatched elsewhere in the world, contract manufacturers in the notebook PC industry have gained more bargaining power against the brand marketers. This creates room for local governments to intervene as power brokers, because part of the power attributed to contract manufacturers comes from the support of local governments. While the local governments have found room to maneuver to influence the location decisions of related firms in the

GPNs, they have also found it difficult to replicate the labor market institutions found elsewhere that underscore the strengths of the GPNs. The relocation is unlikely to succeed if the local government in the new location does not come up with some institutional innovations to reshape the structures of the GPNs.

The information used in this paper is mainly collected from a series of interviews conducted in Sichuan Province, where a couple of notebook PC clusters are emerging. Top executives of the world's major notebook PC contract manufacturers in Chongqing and Chengdu were interviewed in October 2011. This information is combined with the authors' previous interviews in the PC clusters along the East Coast of China over the years, as well as secondary data related to the individual firms or the notebook PC industry as a whole.

A Case Study

This study looks at the relocation of Taiwanese-owned contract manufacturers (contractors for short) of notebook PCs from the Shanghai-Suzhou region of China to the western province of Sichuan. The attempted relocation began in 2009, after the devastating 2008 Subprime Financial Crisis that left millions of migrant workers in urban China unemployed (Chan 2010). Migrant workers formed the mainstay of the workforce of Taiwanese PC contractors in the Shanghai-Suzhou region. They mostly migrated from the inner provinces of China, notably Sichuan, Hupei, and Hunan, to work in urban factories located in the coastal areas. The number of migrant workers, defined by the Chinese statistical authority as those who work outside the territory of their household registration, was estimated to be 253 million in 2011 (China Bureau of Statistics 2012).

Household registration (*hukou*) is a unique Chinese institution, which serves as a mechanism for population control and the distribution of social benefits. Migrant workers who live and work in the coastal cities receive either none or only part of the social security benefits that are available to local residents. The treatment of migrant workers differs according to location. Typically, migrant workers are denied unemployment compensation and pensions, but they are entitled to partial coverage of health insurance and work injury protection (Wang 2011). In cities where public housing or housing subsidy programs are provided, migrant workers are normally ineligible to receive such benefits. Migrant workers typically live in dormitories built by their employers because they cannot afford private housing. The education of migrants' children used to be denied by host cities, but has recently become more available as a result of a central government mandate. Because the provision of social benefits is strictly tied to the location of the individual's hukou, migrant workers relocate to the city when they are young and return to the countryside when they get old (Fan 2011). While China is not the only country in which uneven development has led to a dislocation of labor (Yeates 2004), no other country has experienced such a massive dislocation of rural residents as has China.

A massive, young labor force has underlain the rapid growth of China's export-processing industries

Female migrant workers, who are most visible in the export-processing factories in coastal areas of China, typically start their "career" at the age of 18–20, and retire as soon as they get married (Pun 2005, 5). This specific pattern of labor supply has created a massive, young labor force that has underlain the rapid growth of China's export-processing industries since the economic reforms began in 1979. The coastal areas can call upon this seemingly unlimited reserve army for regional development by paying only a fraction of the labor reproduction costs (Burawoy 1976).

It is ironic that the immobility of labor within China under the hukou system makes migrant labor a common asset for the coastal areas, which are relatively attractive to MNCs because of geographical advantages or policy preferences. Cities in the coastal areas compete to lure foreign capital, treating migrant labor as a public good in the race. With the steady availability of migrant workers, the local labor force becomes inconsequential in the competition for foreign capital. Instead, connections to major players in the GPNs are more important among local competitors vying to attract export processing industries. The Shanghai-Suzhou region succeeded in creating a world-class information industry cluster mainly because of its connections to MNCs, particularly Taiwan-based contract manufacturers in the industry. The region provided many friendly institutions that have enabled Taiwanese contractors to successfully compete in world markets. These include large transport capacities, efficient and accommodative customs procedures, and favorable tax treatments. Whereas unskilled workers are inconsequential to regional competition, skilled workers are crucial to interregional rivalry. Many urban cities in the coastal areas provide solid programs to attract and retain skilled workers. For example, the city of Kunshan, where a cluster for the notebook PC industry was established, offered a generous migration policy for out-of-region skilled workers who sought local resident status, and a special allowance for Taiwanese investors to operate a Taiwan school to educate the children of expatriates. It is believed that the close alliance between the local government and the Taiwanese investors explains the success of the Kunshan cluster (Wu 1997).

This unique labor market condition affected the power structures of the GPNs that were embedded in local institutions. Aided by a seemingly unlimited supply of migrant workers, Taiwanese investors expanded their market shares in the contract manufacturing business of notebook PCs. Five leading contractors—Quanta, Asustek, Foxconn, Compal, and Invetec—together accounted for 90 percent of the world's notebook PC production, with total shipments exceeding 150 million sets in 2011. Moreover, almost all shipments originated in the Shanghai-Suzhou region. The top five contractors served the five leading brands of Hewlett-Packard (HP), Acer, Dell, Lenovo, and Asus. The relationships are interwoven, with each brand working with two to three contactors and, conversely, each contractor serving two to three brands. While each brand is identified with one major contractor, the runner-up contractor follows closely in terms of its business share. These two contractors are considered to be strategic partners, responsible for both product development and manufacturing the products. The distant third contractor mainly serves as an alternative source of supply and an insurance against contingency. This structure dramatically differs from that of the past, when China was still absent from the global production of notebook PCs. In former days, the number of brands was similar to what it is today, although with different names, but the number of contractors was much larger. The brands used to completely dominate the GPNs. In the early 2000s, Taiwanese contractors were forced by the brands to relocate their production to China and congregate in the Shanghai-Suzhou region, despite the objections of the Taiwan government (Chen and Ku 2012).

The new structure of the GPNs that has evolved in China has brought the relationship between brands and contractors closer to being an alliance, rather than an arms-length transaction partnership. The contractors perform more tasks on the value chains than they did in Taiwan. They codevelop products with the brands, manufacture the products, and ship and store them in warehouses around the world before delivering them to stores or customers designated by the brands. Contractors are now sometimes called "global logistics" service providers. To further reduce the coordination costs in the GPNs, the contractors have increased the degree of vertical integration and encouraged their first-tier suppliers to do the same. For the notebook PC industry, the first-tier suppliers include those making outer cases, motherboards (of electric circuits), and keyboards. As a result of market concentration in contract manufacturing, first-tier suppliers have become even more concentrated than contractors, as their production is very capital-intensive. For example, in China there are only two major keyboard suppliers and three outer case suppliers that serve all contractors in the Shanghai-Suzhou region.

This new structure, which has been created with the ultimate purpose of cutting transaction costs, gives contractors much more bargaining power vis-à-vis the brand marketers than in the past. Although the brands continue to push the contractors to cut costs every year, the contractors can also partially pass on cost increases attributable to local conditions, such as labor costs, to the brands (Appelbaum 2008). The fact that the number of contractors is small secures their bargaining position. Furthermore, since they are all located in the same region and subject to the same local conditions, it is difficult for the brands to play one contractor against the other in price bargaining. Being bound by the same cost structures also increases the collective bargaining power of contractors against their clients. More importantly, these cost structures can be manipulated by local governments, and, when desirable, governments can alter the local conditions to tilt the power in favor of the contractors. In a nutshell, the contractors are strategic partners with local governments, while the brands are not.

> *Contractors are strategic partners with local governments, while the brands are not*

It has been argued that, in global competition, the role of the local government is to create local institutions that reinforce the capacity of international investors to exploit spatial differences (Yeung 1998). In China, the most manipulated local institutions are labor institutions. Underlying the production clusters in the coastal areas of China are millions of migrant workers who live outside their hometowns, often located thousands of kilometers away. Most of them live on the corporate compounds and spend their leisure time using the recreation facilities provided by the company. They are willing to work overtime and on holidays in order to earn bonus wages, as their length of employment is often quite short. Despite government regulations, many work beyond the legal limits. They can be hired and discharged at a minimum cost as there is an enormous pool of rural labor waiting to flow to the cities. This gives employers great flexibility in production arrangements. When there is a need to rush production in the short run, migrant workers are ever ready to work extra hours. When there is a need to increase the workforce in the long run, there is a seemingly unlimited supply of available people.

However, the turnover rate of migrant workers is very high, ranging from 30 to 40 percent a year. In fact, migrant workers normally go back to their hometowns during the Chinese new year holidays, and only about 80 percent are expected to return to their original work posts. In addition, there is significant turnover during the regular seasons. In response to this, contractors have modified their production technologies to reduce the skill requirements of shop-floor workers. In most cases, training for new recruits to perform shop-floor tasks does not exceed one week. This makes the replacement of exiting workers with new recruits a seamless process. For skilled jobs, the contractors prefer local workers, whose turnover rate is much lower. Local governments obligingly accommodate this division of labor by offering a friendly policy environment for skilled migrant workers to become local residents. For example, in Kunshan, the local government offers resident status to anyone who purchases a property in the city. The employer can also purchase property on behalf of migrant workers (without transferring ownership), enabling them to obtain resident status if their skills are desired. Essentially, what this policy does is to create a dichotomy in the labor market, one for local workers and one for migrant workers. Migrant workers are discriminated against in terms of the social welfare attached to their work. Their status can only be changed by a selection mechanism controlled by the local governments in consultation with the MNCs. The population of Kunshan today is estimated to be 1.7 million, of which only 600,000 are local residents; the rest are migrants with temporary status based on the hukou.

With this dichotomized labor regime, the production of the MNCs is anchored by local-bound, skilled workers, whose pooling is considered to be an indispensable factor in the formation of an industry cluster (Marshall 1920; Bresnahan, Gambardella, and Saxenian 2001). Whereas unskilled workers come and go, skilled workers are here to stay, although they may move between firms in the cluster. The localized mobility of skilled labor is believed to be the major source of external benefits in a cluster (Eriksson and Lindgren 2009). Skilled workers are a location-specific asset, and their existence defines the boundary of a cluster. They assume the role of knowledge accumulation and the responsibility of transferring this knowledge to unskilled workers during on-the-job training. They are local residents by birth or by "naturalization." In contrast to unskilled workers who are in flux, skilled workers are location-bound. When skilled workers are deployed to a new location within China, their costs increase because employers have to compensate them for the losses of social benefits that are tied to their original residency. Their status is similar to that of expatriates dispatched by MNCs to foreign countries. This makes relocation of production within China just as difficult as relocating to a foreign country.

Meanwhile, the vertical integration of production has bundled the skill-intensive and the labor-intensive processes together. Relocating labor-intensive processes alone to a new location requires a disintegration of production and a reconfiguration of the GPNs. It is not a simple change of production location, but a reengineering of the supply chains. Even if this can be done, the advantages of having a single location for both production processes and access to skilled workers may be lost forever. In the case of China, immense transport costs are also incurred by the vertical fragmentation of the production processes. Moreover, when unskilled jobs previously performed by migrant workers are instead performed by local workers, the flexibility intrinsic to migrant workers is lost. Local workers go home after work rather than staying in the dormitories. The employers and the local state together have to bear the entire labor reproduction costs. Local laborers have less incentive to work during off hours, not to mention on weekends and holidays. Therefore, only production with a predictable schedule will fit the circumstances. The result is a loss of the flexible production that has made the East Coast of China so competitive.

Westward Movement

The Subprime Financial Crisis of 2008 sent millions of migrant workers home as China's export industry suffered a drastic decline in world demand (Chan 2010). When the export orders began to rebound in the second half of 2009, only a portion of the migrant workers returned to the coastal areas. When the economy was fully recovered in 2010, widespread labor shortages were reported in provinces that host export processing industries. Many scholars believe that the persistent labor shortages in China were a manifestation of the Lewis turning point, which signals the disappearance of surplus labor (Cai and Du 2011). Around this time, the notebook PC contractors on the East Coast attempted to relocate to Sichuan Province, prompted by local government incentives. By 2011, all five major contractors had established some operations in Chongqing and Chengdu, two major cities in the province of Sichuan. However, they encountered serious problems in trying to replicate what they had been doing in the coastal areas. These problems will be discussed later in this paper.

When the economy fully recovered in 2010, labor shortages were reported

The reason why the contractors chose Sichuan over other inner provinces, all of which had been promoted by the central government as suitable destinations for Chinese industries moving from the coastal areas, was the strong policy initiatives offered by the local governments, in addition to the fact that Sichuan had the largest internal market among the western provinces. The local governments, especially those of Chongqing and Chengdu, thoroughly understood the supply chains of the PC industry, and they worked sequentially on brands, contractors, first-tier suppliers, second-tier suppliers, and so on to engineer a collective relocation of the industry. They first offered incentives to brand marketers to encourage them to relocate their regional operations centers to Sichuan, including concessions on the business tax and price subsidies on the products sold in the jurisdictions. The central government allows MNCs located in the western provinces to enjoy a reduced corporate tax rate of 15 percent, as opposed to a normal rate of 25 percent applied elsewhere in the country. The 15 percent tax is composed of a 9 percent tax accruing to the central government and a 6 percent tax accruing to the local government. The local government typically makes a concession on the 6 percent tax that goes into its fiscal coffer. For example, it may offer a period (say, two years) of tax holiday for this portion of the tax, followed by a period (say, three years) with a reduced rate (say, 3 percent). In addition, the local government offers price subsidies on the notebook computers sold in local markets. The subsidies normally take the form of rebates of the value added tax (VAT). VAT is 17 percent, and a quarter of the revenue is shared by the local government. The rebates often go beyond the local share of 4.25 percent and can go as far as 10 percent.

Strong tax and fiscal incentives lured HP, Acer, and Asus to set up regional operations centers in Chongqing, and Dell and Lenovo in Chengdu. The VAT tax rebates went to brand marketers. After succeeding in recruiting the brands to their cities, two local governments began to work on major contractors of these brands. In the case of Chongqing, HP was the first to be on board, followed by three of its contractors. Acer was the second to be on board, followed by two of its major contractors. Asus was the third, which came to share the contractors that had already relocated to the region. It was not surprising that the HP and Acer groups chose to locate in two different industrial estates in Chongqing. The local governments offered the contractors concessions on corporate income tax, plus a full subsidy on the transportation costs incurred while shipping materials and parts from the coastal areas that were not yet available in the region. The subsidy was to last for three years, as the local governments believed that major suppliers would soon follow in the footsteps of the contractors in relocating to Sichuan. It also subsidized the transportation costs of shipping finished computers that exceeded

> *Strong tax and fiscal incentives lured HP, Acer, and Asus to set up regional operations centers in Chongqing*

the costs of a typical shipment from the East Coast. Transportation costs were one big disadvantage for export-oriented operations locating away from the coast, despite the great push in recent years to build highways, railroads, and airports connecting inland cities. The entire logistics system in China remained inefficient. It was reported that total logistics costs accounted for 17.8 percent of China's GDP in 2011, twice that of other developed countries (*People's Daily Online* 2012). In addition to the tax concessions and transport subsidies, contractors were offered spacious land on which to build factories at very low prices. Local governments could also build and lease the factories to the contractors, if the contractors preferred that option. The local government of Chongqing even negotiated with major brands to offer contractors extra fees for the products manufactured in the city. The extra fee for each notebook computer was said to be US$5, which was more than 5 percent of the typical manufacturing fee that the contractors would receive from the brands.[1] The local government had this leverage because it offered price subsidies to the brands on computers that were sold locally. In order to qualify for the subsidy, one locally sold computer had to be matched by one computer exported. In other words, the local government took away US$10 from each subsidy that it offered the branded companies and transferred it to the contractors.

After the relocation of the contractors, the local government began to work on the first-tier suppliers, focusing on makers of outer cases and keyboards, for which proximity to final assembly is critical to production efficiency. These makers were also offered tax concessions and low-priced land, but no subsidies on transportation. By the end of 2011, two keyboard producers and three outer case producers had commenced production in the Chongqing region. Meanwhile, city officials said that a list of more than 500 second-tier components and parts suppliers in the Shanghai-Suzhou cluster were being targeted, and many of them had already agreed to invest in the two cities (*United Daily News* 2012). City officials of both Chongqing and Chengdu had a vision to replicate a notebook computer cluster similar to that in the Shanghai-Suzhou area. For example, the city government of Chongqing envisioned a cluster capable of making 50 million notebook PCs each year. In 2011, the total shipment of notebook PCs from the Shanghai-Suzhou region was estimated to be 150 million units. Officials believed that 50 million units were enough to support a full-blown PC cluster with most parts locally sourced; a few key components are monopolized and locked in specific locations, including the CPU, memory chips, and disk drives. In addition to notebook PCs, other information products such as printers and monitors may also move to the cluster if the PC supply chains become more complete.

> *Chongqing envisioned a cluster capable of making 50 million notebook PCs a year*

Labor Market Conditions

Although the relocation of supply chains has shown some concrete results, as evidenced by the total shipment of notebook computers from Chongqing alone reaching 20 million units in 2011 (*Wenweipo* 2012), the core problem of labor shortages has continued to haunt the contractors, and may eventually limit the development of the cluster. All three contractors that commenced full operations in Chongqing reported serious labor shortages. Each employed 30,000 to 50,000 workers in the Shanghai-Suzhou region, but they were struggling to find 10,000 workers for the initial stage of the Sichuan operations. Chongqing city officials had to order township governments within the city jurisdiction to recruit and send workers to the designated industrial estates to meet the demand. The city government offered each migrant worker relocating from the rural villages and towns a 1,500 RMB moving subsidy, provided that he or she worked in the industrial estates for more than three months. The subsidy was equivalent to about two months of wages. The city government also built dormitories and low-priced rental apartments to host the migrant workers at subsidized rates, an action rarely seen along the East Coast. In other words, the local government shouldered part of the reproduction costs of labor (Burawoy 1976) that were typically borne by East Coast employers.

However, the labor shortages remained. The key to this problem is that cities like Chongqing and Chengdu, despite being major metropolises in Western China, command a much smaller pool of the migrant labor force compared to major cities on the East Coast. Migrant workers are motivated by the wage differential between their hometowns and the host city, and the wage differentials offered by Chongqing and

Chengdu were much smaller than those in the coastal cities. For example, in 2011, the minimum wage in Chongqing was 870 RMB, compared to 1,280 RMB in Shanghai and 1,500 RMB in Shenzhen. Whereas Shanghai and Shenzhen were able to attract migrant workers from all provinces in the country, Chongqing was attractive only to the neighboring provinces. In other words, the "backyard" from which migrant labor could be drawn was much smaller.

Contractors in Chongqing reported that the majority of their workers came from Sichuan Province, and only a minority was out-of-province labor. For the in-province workers, most lived more than two hours away in terms of travel distance. In other words, they could not commute to work and had to live in factory dormitories or low-priced apartments provided by the local government, but they could go back home during weekends or holidays.[2] Compared to the extra-province migrant workers, intra-province workers were generally older and more likely to be married. The 2011 statistics (China Bureau of Statistics 2012) on migrant workers showed that among the extra-province migrant workers, only 18.2 percent were aged 40 or older, while this ratio was 60.4 percent for intra-province migrant workers. While 58.2 percent of extra-province migrant workers were married, the ratio was 90.2 percent for intra-province migrant workers. It is only natural for older, married workers to be emotionally bound with their families, and to want to go home whenever they can. While working closer to home is good in terms of reducing emotional stress, which is believed to be the major cause of a series of suicide incidents in mass factory compounds like those of Foxconn in Shenzhen (Fair Labor Association 2012), being closer to home reduces the incentive to work during off-hours. This undermines the flexibility of production scheduling that is critical to the competitiveness of contractors. At the present time, all contractors interviewed for this paper ran a one-shift production schedule, although they all wished to switch to a two-shift schedule as they had done on the East Coast. The inflexibility of labor, together with the need to haul selected components and parts on a two-day truck journey from the coastal areas, prevent the contractors from replicating the "just-in-time" production method of the East Coast. The "just-in-time" scheme is essential to multiple-model production, which caters to market demand on a "built-to-order" basis. It is typical that major PC brands like HP offer 40 to 60 models per year through their versatile contractors.[3] At the present time, production in the factories in Chongqing and Chengdu is confined to mature models, particularly consumer-oriented (as opposed to commercial) models, with predictable quantities of sales. In other words, they produce according to forecasted demand, in the old-fashioned way before the "built-to-order" business model came to dominate the industry.

Although most employees are migrant workers based on the official definition, they are within-province workers. It is more difficult to shortchange within-province migrant workers of their social welfare entitlements, as most social security programs are managed by provincial governments. The result is an increase in social security contributions on the part of employers compared to the factories in coastal cities. The local government cannot deny social welfare coverage for these "migrant" workers either. In other words, unlike on the East Coast, employers and local states in Sichuan have to cover the entire cost of labor reproduction. For example, in spite of the fact that the minimum wage in Chongqing was only two-thirds that of Shanghai, the contractors reported that the total labor cost, inclusive of social security contributions, amounted to 85 percent of Shanghai's costs.[4]

An important reward for their generous social welfare contribution, however, is that the labor turnover rate is much lower than on the East Coast. In view of this reality, the local government has been trying to create institutions that reinforce this advantage for investors. For example, the government of Chongqing initiated a policy to increase the labor force in the city's urban areas by offering some of the most generous conditions in the country for migrant workers to obtain residential status (McNally 2004). In 2010 and 2011, the Chongqing government reported that 3.22 million migrant workers were awarded an urban hukou in Chongqing (*China News Online* 2012), constituting a 10 percent addition to the city's population, which was estimated to be 31 million in 2011.

Adding to the complexity is the need for skilled workers. As mentioned in the previous section, the production technologies of Taiwanese contractors are reliant on skilled workers, and most skilled workers are local natives or residents who are essentially location-bound. The relocation of production lines from the East Coast must be accompanied by a relocation of skilled workers, as they are the carriers of technologies. With their welfare benefits tied to the city where their hukou is located, skilled workers can only be dispatched to new places on a temporary basis. Their deployment to Sichuan is much like an assignment to a foreign country, and has to be compensated with extra pay. In fact, many of these "expatriates" were originally Sichuan natives who had migrated to the East Coast, making them far more willing than their colleagues to take assignments back home. However, most of them do not expect their assignments to be permanent.

During the rapid development of China's coastal provinces, most inland provinces experienced a brain drain in terms of the outflow of skilled workers. The brain drain was most serious in Sichuan (Li 2004), where the local government established a policy to encourage the "homecoming" of native skilled workers who had established their hukou in coastal cities. Chongqing, for example, offers returnees the same social benefits as local residents, while allowing them to retain residential status in the coastal cities.

> *During the rapid development of China's coastal provinces, most inland provinces experienced a brain drain*

The consensus among contractors is that the Shanghai-Suzhou PC cluster will be impossible to replicate in Sichuan because of a smaller pool of workers, both skilled and unskilled. The smaller pool is likely to put a constraint on the size of the cluster. In addition to size, the structure of the labor market also matters. On the East Coast, a dichotomized labor market characterized by a very high turnover rate led the contractors to develop a labor management system marked by little training but very stringent shop-floor disciplines. Unskilled workers were considered disposable and easily replenished. Resources were only expended on maintaining skilled workers, who were location-bound. In comparison, the local government in Sichuan was trying to create an integrated labor market by liberalizing the migration policy. Employers were also more willing to expend resources on training and retaining unskilled workers as they became valuable resources. Compared to the East Coast, where unskilled workers typically performed a single task, in Sichuan they were likely to be trained for multiple tasks.

The Power Structure of GPNs

The power structure of GPNs determines the relocation processes. When Taiwanese contractors first relocated from Taiwan to Eastern China, they basically followed the instructions of the brands and were supported by the brands (Chen 2003). When they moved from Eastern China to Sichuan, they were mainly prompted by local government initiatives. The government of Chongqing had obtained the commitment to relocate from at least one major contractor of HP before offering fiscal incentives to HP, which was the first brand to move its operation's headquarters to Chongqing. The local government also bargained on behalf of the contractors for a favorable price for their manufacturing services in Chongqing. All of these are indicative of the enhanced bargaining power of contractors vis-à-vis the brands in the GPNs, as well as a strategic coupling between the local state and the contractors. The only major brand that has until now refused to allow its computers to be manufactured in Western China is Apple, which still maintains a dominant position in the GPNs.

Power relations in the GPNs can be explained by the resource dependency within the network (Pfeffer 1981). Those who control critical resources in the GPNs obtain power in relation to those who depend on these resources. Taiwanese contractors gained more power after they relocated from Taiwan to China because they increased their world market shares in the contract manufacturing business and because they formed a closer alliance with their suppliers. This alliance allows contractors to bridge the structural holes in the GPNs, which confers power in the networks (Burt 1992).

The alliance relationship between contractors and the first-tier suppliers is most illuminating. As the contract manufacturing business in the PC industry became more concentrated, the market for key components and parts also became more concentrated. A small number of players in the industry have laid the foundation for alliances in lieu of arms-length transactions. This is most evident in the case of outer case makers. Outer cases vary with each model of computer, and they need to be produced in proximity to the final assembly lines to enhance the efficiency of coordination. Good coordination between the contractors, who design and conduct the final assembly of computers, improves the quality of products and reduces the risk of defects. Collaboration jointly benefits the contractors and casing suppliers.

Along with forging strong alliances with the contractors, casing production has become more vertically integrated as the scale of production has increased. Before relocating to Eastern China, the casing production was divided into three distinct processes: molding (forming and shaping), coating (surface treatment), and assembly, each performed by independent firms. After a few years of evolution in Eastern China, all these processes were integrated within one company. Both molding and coating processes are very capital-intensive, especially when metal cases are produced. Compared to contractors, outer case producers relocating to Sichuan require much larger amounts of capital, and expose themselves to greater risks. Therefore, the contractors did not move without the consent of their outer case partners. In fact, the decisions to move to Chongqing and Chengdu by three major contractors were made concertedly with two major suppliers of plastic computer cases.[5] The other two contractors have their own subsidiaries making plastic computer cases, and decisions were also made jointly. Because of the nature of capital-intensive production, outer case producers are less sensitive to labor costs compared to the contractors, and hence are less motivated to move. It was the contractors who took the initiative to relocate and who persuaded their outer case partners to join them. Although they have remained separate entities, the contractors and outer case producers consider their relationship to be one of virtual integration. The relationships between the contractors and the keyboard and motherboard producers are similar, although not as intimate.

The contractors and outer case producers consider their relationship to be one of virtual integration

The vertical integration of casing production has bundled the capital-intensive process with the labor-intensive process. If the labor-intensive process is more footloose, as the literature suggests, the fact that it can no longer be decoupled from the capital-intensive process makes the industry as a whole less mobile. Mobility is restrained because the commitment of large amounts of capital entails a high risk. So far, only plastic casing production has relocated to Sichuan. The production of metal casing remains in the Shanghai-Suzhou areas, with their manufacturers adopting a wait-and-see position. Metal-based computer cases are even more capital-intensive than plastic ones. Two keyboard producers have relocated to the Chongqing and Chengdu areas, but the metal parts used in keyboards are still shipped from the East Coast.

In addition to risk-sensitive capital expenditures, operating capital equipment usually relies on skilled workers, whose limited availability in the inner provinces further confines the mobility of capital-intensive operations. In fact, even finding unskilled labor can be a constraint. Vertical integration makes scale critical to the efficiency of production. When labor is in short supply, bottlenecks arise in the labor-intensive production processes, making it difficult to achieve scale economies. This forces the producers to ship from the East Coast those components and parts that can only be economically produced in large volumes. Doing so eliminates the benefits of vertical integration, which becomes a disincentive for relocation. The fact that process-specific parts, such as casing and keyboards, are produced in an integrated way contributes to the immobility of the so-called footloose industry. In other words, process-specific investments, which serve to reduce the transaction costs for the GPNs located on the East Coast, also become an entry barrier for the other regions.

Compared to process-specific parts such as outer cases and keyboards, process-independent parts and components such as CPUs, memory, and disk drives are less relevant to the relocation decisions. Process-independent parts can be produced separately in faraway locations. For example, CPUs are designed and fabricated in the United States and assembled in Malaysia (with some in Dalian and Chengdu); memory is chiefly manufactured in South Korea and Japan; and disk drives are mostly made in Thailand, with some in Wuxi (Jiangsu Province), China. These components are stand-alone devices, insensitive to the assembly processes, and are therefore immune to the movement of global factories. They form the core of modularized production in the PC industry (Baldwin and Clark 2000). When the global factories relocated from the East Coast to Sichuan, these components continued to be produced in their original locations. The suppliers only set up VMI (vendor managed inventory) sites close to the new factories in order to offer timely delivery of the parts. The suppliers are mostly monopolists in their specific fields, and they deal directly with the brands rather than the contractors. Orders are made by the brands, and products are delivered to the factories on a consignment basis. In other words, process-independent parts are largely unaffected by the local conditions that determine the geography of global factories. Despite the efforts of local governments to attract CPU production to China, so far only Intel has set up assembly and auxiliary operations in Chengdu and Dalian. The fact that these parts are modularized makes the relocation of the downstream assembly operations relatively easy. They are, in fact, an important contributor to the footloose industry.

Policy Implications

The relocation of PC manufacturing to the coastal areas of China in the last 20 years has had a profound impact on the structure of GPNs in the industry, making the contract manufacturers less footloose than before. Two major phenomena can be observed: one is that the number of players in the GPNs has been significantly reduced, while the other is that the GPNs are deeply embedded in local institutions. As the number of contractors and their first-tier suppliers has decreased, the power structure of the GPNs has shifted in favor of the contractors. Along with the shrinking number of players, production has become more vertically integrated, and the alliances between the contractors and their chief suppliers have been strengthened. The contractors have become more assertive in making location decisions, and the decisions have tended to be made jointly with their chief suppliers. Meanwhile, as GPNs are deeply embedded in local institutions, local governments have become strategic partners with the contractors. Relocation entails a search for the right local government partner. That partner is willing and able to create the right local institutions, which can best accommodate the GPN structures that have already adapted to their current locations. Because a perfect replication of local institutions is difficult, a wholesale relocation of GPNs from one place to another is also difficult. The concept of a footloose industry, which implies the perfect mobility of the said industry, needs to be qualified.

To the extent that local institutions matter to the geographical location of global production, local government has an important role to play in attracting global firms. Local institutions must be shaped in such a way that the local advantage reinforces the competitive advantage of global firms. It is erroneous to assume that labor is homogeneous across nations and is, therefore, dispensable at the discretion of global firms. In fact, it is more accurate to treat labor in combination with its related institutions as a package in global competition. When labor institutions are taken into account, labor is no longer homogeneous, but is more likely to be country-specific. Country-specific factors are essential to the competitiveness of global firms (Dunning 1998), and they are incompatible with the footloose industry, which is based on the assumption of homogeneous factors.

When studying the effects of labor institutions on the geographical location of global firms, most attention has been directed to labor market regulations, such as labor standards, minimum wages, labor unions, etc. (Riisgaard and Hammer 2011). This case study, however, suggests that it is important to look at the reproduction costs of labor. Labor reproduction costs include maintenance costs and renewal costs (Burawoy 1976). Maintenance costs refer to the costs of keeping workers productive while they are on the job. These costs include subsistence wages, work injury protection, health insurance, and so on. Renewal costs refer to the costs of replacing workers who continuously exit the labor market. These include costs incurred in childbirth, child care, education, and training. Some social welfare benefits such as housing, unemployment, and pension programs serve both maintenance and renewal purposes. Labor reproduction costs, which can be regarded as the entire social cost of keeping a capitalist system functioning, are typically shared between employers and society. The distribution between the two is determined by labor institutions. Wages, which account for only a fraction of the labor reproduction costs, do not accurately reflect the competitiveness of the region in global production.

This case study suggests that it is important to look at the reproduction costs of labor

China has a unique labor institution in its hukou system, which allows employers in the coastal areas to bear only a small share of the labor reproduction costs. Most of these costs are shifted to the localities from which the migrant workers have originated. Global firms located in the coastal areas not only benefit from low wages, but also by avoiding most nonwage portions of the labor reproduction costs.

This advantage exists as long as migrant workers continue to flow into the region. Therefore, for employers in the coastal areas, the disappearance of migrant workers—or, synonymously, the coming of the Lewis turning point—means not only a rapid increase in wages, but also an increasing share of labor reproduction costs. That is, they have to worry about both the maintenance of labor and the renewal of labor. Because of the unique labor institutions in China, the impact of the Lewis turning point on the competitiveness of export industries in the coastal areas will be stronger than its past impact on other developing countries, such as those in East Asia.

It is interesting to note that the hukou system, which was designed to restrain internal migration, turned out to be an institution that enabled the coastal provinces to exploit the inner provinces. The hukou system coupled well with China's strategy of pursuing an uneven regional development, with the coastal areas developed first. The coupling resulted in very rapid growth in the coastal areas, but it also exacerbated the disparity between regions. During this process, multinational corporations (MNCs) have been important vehicles in implementing the policy, as MNCs were keen to take advantage of regional disparities. In a sense, MNCs were strategic partners with China's uneven development policy. When China repealed its uneven development policy in 2000 and began to promote the westward movement of industry, the question arose as to whether MNCs would still be strategic partners. The apparent answer is no, as MNCs have not shown any interest in helping China's government eradicate its regional disparities. They could only be motivated to move westward through fiscal incentives, such as the subsidies offered by local governments of Sichuan. Therefore, compared to the coastal areas, the ability of the inner provinces of China to mobilize MNCs to aid their industrial development has been quite limited. They have had even less success in mobilizing export-oriented MNCs because of the additional transportation costs incurred when production takes place in the inland areas. In other words, it is difficult for the inner provinces of China to replicate the success story of the coastal area's MNC-aided industrial development. Whereas in the coastal areas the local states and MNCs joined hands to exploit the resources of the inner provinces, the local states in the inner provinces have had to subsidize the MNCs to gain their favor.

However, there is no doubt that the coastal areas of China have lost their location advantage in hosting the global factories because of changing labor market conditions. The institutions underscoring this advantage have become irrelevant at the same time, and a new set of institutions can be constructed elsewhere to reshape the nature of global competition. Indeed, competition among local institutions has been a very important part of the regional rivalry in China since the economic reforms began in 1978 (Xu 2011). Regional competition leads to innovations in institutions, which may reshape the future course of global competition.

The literature has suggested that to succeed in global competition, it is possible to have multiple solutions for bundling local institutions with GPN structures (Yang and Coe 2009). The key is for local institutions to reinforce the comparative advantage of the region, which in turn gives a competitive edge to the GPN players located there. Because the comparative advantage of each region differs, the institutions created are also idiosyncratic. The labor institutions created by the local government in the Shanghai-Suzhou region have enabled the notebook PC brands to offer "built-to-order" services, which have been a destructive innovation in the global PC market. It is hard for the rest of the world, and even the rest of China, to replicate this system because no other region can amass such huge numbers of migrant workers that can be hired and fired with ease. If Sichuan Province is to succeed in establishing a new cluster of notebook computer manufacturers in the future, the labor institutions created there would necessarily be distinctive from those along the East Coast, and the service model offered by the region would also likely be different. In other words, a new labor institution would have to be matched with a new GPN structure, which, in turn, would offer a new kind of service model to the global market.

The major difference between Sichuan Province and the coastal areas is the nature of the migrant workers. Migrant workers who used to be present in the coastal areas were young, unmarried, and characterized by a very high turnover rate. MNCs in the coastal areas organized their production to accommodate the characteristics of this labor force. In essence, the production process was de-skilled so as to minimize the job requirements, and the production scheduling was made extremely flexible. This production organization allowed the notebook PC industry to offer a unique service model that outcompeted rivals in other countries. When these MNCs relocated to Sichuan, migrant workers were fewer in number, older in age, mostly married, and less inclined to leave. Production had to be reorganized to suit the distinctive nature of the labor market there. More importantly, the local government had to restructure the labor institutions in such

> *The major difference between Sichuan Province and the coastal areas is the nature of the migrant workers*

a way that the strengths of the local labor market were reinforced. The Chongqing government's adoption of a liberal migration policy to increase the thickness of the labor market and to reduce the turnover rate was an effort in that direction.

In fact, the labor institutions that the local government in Sichuan had attempted to create were converging with models established in the rest of the world. Despite losing the advantage of migrant workers, Sichuan has not been rendered uncompetitive in hosting global factories. The phenomenon of vast numbers of migrant workers may have been unique to China, but it has tended to be transitory. As in other developing countries, surplus labor eventually disappears as the economy develops. Given the fact that no other country can replicate the kind of massive internal migration that has occurred in China, Sichuan can actually compete with other countries on a "leveled playing field." In leveled competition, the quality of labor determines the competitiveness of a region. Traditionally, high-quality labor is characterized by low turnover, high education levels, harmonious labor relations, and so on. It is also apparent that a good social security program is indispensable to the maintenance of high-quality labor. Unlike in the coastal areas where differentiated treatments are offered to workers based on their resident status, it is probably more productive to provide universal social security coverage to all employees in the region. The literature also points out that labor institutions that facilitate the pooling of skilled workers are essential to the formation of an industrial cluster (Marshall 1920; Bresnahan, Gambardella, and Saxenian 2001). This suggests that the labor institution must be conducive to the congregation of skilled labor in the region.

On the other hand, as labor market conditions have drastically changed in the coastal areas of China, the institutions that have underscored the region's comparative advantage have to be modified as well. Local institutions that fail to adapt to changing market conditions or technological revolutions become obsolete, and the region in question becomes stagnant (Wei, Li, and Wang 2007). Institutions, however, are often slow to adapt. Faced with a decreasing supply of migrant workers, many cities in China's coastal areas have repeatedly raised their minimum wage since 2010 (*Financial Times*, 2012). This policy does not seem to have eased the labor shortage problems, however. Some cities have initiated an industrial restructuring program, nicknamed *teng-long-huan-niao* (emptying the bird cage for new birds), which is aimed at attracting new foreign investments that are less dependent on low-cost labor (Asia Business Council 2011). There is also a movement toward factory automation to reduce the dependency on unskilled workers. For example, Foxconn, one of the largest contractors, has announced its intention to increase the number of industrial robots in its Chinese factories to one million by 2013, which will match its employment figure of one million Chinese workers. At the time of this announcement in 2011, Foxconn had about 100,000 robots installed in its factory sites (Hille 2011). Robots can be used to perform tasks such as spraying and wielding that are currently undertaken by unskilled workers.

The real challenge to local governments in the coastal areas of China is whether they can restructure their labor institutions to sustain the growth momentum. Given the fact that the structure of their GPNs is path-dependent and deeply embedded in local institutions, and that it is probably hard for other regions to replicate this structure, an effective strategy may be to build new strengths in the region based on this location-bound structure. For example, the production of notebook PCs has become vertically integrated, and contract manufacturers have gained power in the GPNs. Instead of attempting to attract new investments that promise to use fewer workers for the same value added, an easier approach may be to encourage existing manufacturers to restructure their production to align with new labor market conditions. Upgrading can be accomplished through internal transformation. Contract manufacturers, for example, might redefine the division of labor by outsourcing the labor-intensive parts of production to neighboring areas where labor costs are still low, while maintaining the capital-intensive processes in the region. A new labor institution could be created to accommodate capital-intensive production, which is often skill-intensive as well. This institution would benefit the training, acquisition, and retention of skilled workers. It should also encourage the movement of labor between the coastal regions and neighboring areas. This implies that social security benefits should be able to cross the borders of the region. In short, the boundaries of the industrial cluster could be extended through institutional reforms.

The concentration of a large number of unskilled workers in the coastal areas of China is likely to be a transitory phenomenon in history, and unlikely to be replicated in other regions of China or other parts of the world. This enormous pool of unskilled labor, coupled with China's unique migration policy, has allowed China to become the "world factory."

With changing labor conditions in China, global manufacturing activities will start to disperse from the coastal regions

This world factory has been able to provide most manufactured goods to the global markets, sometimes from a single location. It has led to an unprecedented geographical concentration of manufacturing activities. For example, about 90 percent of the world's notebook computers are shipped from the Shanghai-Suzhou region of China. But with changing labor conditions in China, it is almost certain that geographical concentration at this extreme level will come to an end, and that global manufacturing activities will start to disperse from the coastal regions of China.

The question is, to what extent will the dispersion of global production benefit other developing countries? Scholars have long been concerned about the possibility that China's participation in global production crowds out the potential opportunities of other developing countries (Giovannetti and Sanfilippo 2009). If the crowding-out effect only emanates from low labor costs, then this effect will certainly be a short-term phenomenon. When labor costs rise in China, as is happening now, the other countries will regain opportunities. This is what the footloose industry argument would have predicted. However, this case study suggests that China may have permanently changed the structures of GPNs, making the relevant industries unable to move, and therefore permanently crowding out the other countries.

China's major impact on the structure of GPNs is that the number of players has decreased, while the scale of production of each player has increased. The advantage of a small number of players translates into a reduction of coordination costs and an increased speed of product development, both of which are essential to global competition. If the industry in question is to disperse from its current location, then multisite production is the only solution, as no other country or region can replicate the kind of industry concentration that we have seen in the past. However, the advantage of small-number bargaining remains. The ability to deliver a large quantity of products synchronously around the world remains a must for global brands. This means that brand marketers will continue to prefer dealing with a small number of contract manufacturers, which, in turn, will be responsible for managing the multisite production. The implication of this is that contract manufacturers will become more specialized, and they will become indispensable partners for brand marketers. The power relations in the GPNs will continue to shift in favor of contract manufacturers in spite of production dispersion.

Whether countries outside of China can attract some of the global factories depends on whether they can offer the right labor institutions that accommodate the new structures of global production. These labor institutions must deliver a labor supply that promises to be large, stable, and inexpensive for a reasonable period of time. "A reasonable period of time" is pivotal to these conditions because the relocation of global factories from China incurs higher costs than what was experienced by the so-called footloose industries in the past, and these costs can only be recovered over a sustained period of time. Relocation costs are higher because capital-intensive processes are also included as a result of vertical integration within the global factories. This means that the long-term trend of labor supply, or the demographic structure, will become an important consideration in the relocation decisions of the global factories. Because of the small number of co-located players in the GPNs, a wholesale relocation, inclusive of the entire supply chain, is likely to be the norm. Without it, the advantage of a cluster will be lost. If no other country can offer a set of favorable conditions to convince the major players in the GPNs that a new cluster, albeit smaller in size, can be established outside of China, the dispersion is likely to take place only within China.

The case of the notebook PC industry may also apply to other footloose industries whose GPNs have undergone similar changes after an extended period of operations in China. A common feature of these GPNs is this: a small number of players dominate each stage of production and cluster in a few locations to produce goods that meet the entire world demand. Most electronics industries seem to share this common trait. The notebook PC case may not apply to industries where a large number of producers are scattered in diverse locations throughout China, such as the textile industry. Despite a large amount of textile exports from China today, the GPN structure of the textile industry has largely remained the same as in the past. Therefore, the industry is likely to remain footloose.

> *The case of the notebook PC industry may also apply to other footloose industries*

Conclusion

Economic geographers have long noticed the spatial unevenness of economic activities and have offered theories to explain this phenomenon. One of the most common explanatory variables in such theories are the local factor market conditions (Martin and Sunley 1996). Local factor market conditions are also considered by Michael Porter (1990) to be one important reason why some countries sustain their competitive advantage in certain industries. Footloose industries, however, are not the kind of industries in which the competitive advantage will remain in one specific country, as Porter has highlighted. On the contrary, footloose industries are the kind in which the regional competitive advantage constantly shifts over time. In theory, the only factor that determines the regional competitive advantage of the footloose industry is the wage rate. It is obvious that this notion of the footloose industry is oversimplified.

The underlying assumption of the footloose industry is that labor is homogeneous and location-bound. A homogeneous factor is also a dispensable factor. Therefore, capital moves around to exploit the differences in labor costs that exist because of uneven economic development. Underdeveloped regions offer low wages to attract the footloose industry, which will stay only if the wages remain low. This assumption is oversimplified because labor does not function in a vacuum. Labor functions within certain institutions that define work standards, compensation, industrial relations, etc., and institutions tend to be location-specific. If we treat labor supply and labor-related institutions as a package in regional competition, then labor is no longer homogeneous. This paper shows that in the notebook PC industry, which is believed to be footloose in the traditional sense, local labor institutions play an important role in determining the competitiveness of the region.

The paper highlights one unique Chinese labor institution, hukou. Hukou has allowed the coastal areas of China to make use of labor residing in other parts of China without paying the full labor reproduction costs. This system has made the coastal areas extremely competitive in labor-intensive manufacturing activities and has engendered several important industrial clusters in the region. These clusters are capable of supplying manufactured goods in a very flexible fashion to meet the needs of the entire world. This capability is supported not only by local labor, but also by the entire labor force of China, with the reproduction cost of labor being borne disproportionately by the inner provinces. Hukou reinforced the uneven development strategy of the Chinese government and enabled the coastal areas to grow at the expense of other regions.

When the number of migrant workers fell, the competitiveness of the coastal areas eroded, both because of rising wages and the increasing burden of nonwage labor reproduction costs. As the labor supply conditions change in China, the labor institutions that have made the coastal region so competitive in the past also have to change, otherwise the clusters in the region will start to crumble. That is to say, local institutions have to evolve over time if their regional competitiveness is to be sustained. How the labor institutions in China should be modified depends on how the direction of industrial restructuring is envisaged. However, it is pretty clear that to remain competitive in the global manufacturing industry, China will need a labor institution that brings more stability, or a lower turnover rate, to the local labor market. This implies greater accommodation rather than exploitation of migrant workers. It also means that the competitive advantage of the region needs to be built on the quality of labor rather than on the quantity of labor.

It is apparent that the geographical concentration of global manufacturing activities in the coastal areas of China will decline in the future. However, it is unclear whether the clusters in the region can be fully or partially relocated to other countries. This study suggests that a complete relocation of the footloose industries away from China is probably difficult to achieve because the underlying GPNs are already tied to unique Chinese institutions, which would be hard to replicate elsewhere. Partial relocation is a more plausible scenario. With rising labor costs in the coastal areas and a possible increase in labor mobility within China if the hukou controls are relaxed, there is no doubt that these industries will disperse. It will be interesting to see whether the dispersion will be confined within China or not, an outcome that will demonstrate whether China has the national advantage in these industries or only a regional advantage. In any event, relocation will entail institutional innovations to match the new structures of GPNs, which are characterized by a small number of players capable of producing large amounts of products synchronously to consumers around the globe. Without the aid of the massive and young labor force that has been unprecedented in history, any other countries or even other regions in China that are interested in hosting the so-called footloose industries will face significant challenges.

Complete relocation of the footloose industries away from China is probably difficult to achieve

Endnotes

1. Personal interview with a contractor, Sichuan Province, October 19, 2011.

2. Personal interview with a contractor, Sichuan Province, October 18, 2011.

3. Personal interview with a contractor, Sichuan Province, October 17, 2011.

4. Personal interview with a contractor, Sichuan Province, October 18, 2011.

5. Personal interview with a case supplier, October 19, 2011.

Bibliography

Applebaum, Richard. 2008. "Giant Transnational Contractors in East Asia: Emerging Trends in Global Supply Chains." *Competition and Change* 12 (1): 69–87.

Asia Business Council. 2011. *Economic Transformation of the Greater Pearl River Delta.* Internal report.

Baldwin, Carliss and Kim Clark. 2000. *Design Rules: The Power of Modularity.* Cambridge, MA: MIT Press.

Baldwin, Richard and Anthony Venables. 2010. "Relocating the Value Chain: Offshoring and Agglomeration in the Global Economy," NBER Working Paper #16611, National Bureau of Economic Research, Cambridge, MA.

Bresnahan, Timothy, Alfonso Gambardella, and AnnaLee Saxenian. 2001. "Old Economy Inputs for New Economy Outcomes: Clusters Formation in the New Silicon Valleys." *Industrial and Corporate Change* 10 (4): 835–860.

Burawoy, Michael. 1976. "The Functions and Reproduction of Migrant Labor: Comparative Material from Southern Africa and the United States." *American Journal of Sociology* 81 (5): 1050–1087.

Burt, Ronald. 1992. *Structural Holes: The Social Structure of Competition.* Cambridge, MA: Harvard University Press.

Cai, Fang and Yang Du. 2011. "Wage Increases, Wage Convergence, and the Lewis Turning Point in China." *China Economic Review* 22 (4): 601–610.

Chan, Kam Wing. 2010. "The Global Financial Crisis and Migrant Workers in China: 'There is No Future as a Labourer; Returning to the Village has No Meaning.'" *International Journal of Urban and Regional Research* 34 (3): 659–677.

Chen, Tain-Jy. 2003. "Network Resources for Internationalization: The Case of Taiwan's Electronics Firms." *Journal of Management Studies* 40 (5): 1107–1130.

Chen, Tain-Jy and Ying-Hua Ku. 2012. "Global Production Networks and the Kunshan ICT Cluster: The Role of Taiwanese MNCs." Forthcoming in *Economic Integration across the Taiwan Strait*, edited by Peter Chow. Cheltenham, UK: Edward Elgar.

China Bureau of Statistics. 2012. *Zhongguo nongmingong diaocha jiance baogao* (Investigative Report on the Monitoring of Agricultural Workers in China), www.chinagate.cn; accessed May 7, 2012.

China News Online. 2012. "Chongqing shizhang Huang Qifan: tuidong nongmingong huji zhidu gaige changtaihua (Mayor Huang Qifan: Making the Household Registration Reform for Migrant Workers a Norm)," January 20, http://www.chinanews.com/gn/2012/01-20/3617994.shtml.

Coase, Ronald. 1937. "The Nature of the Firm." *Economica* 4 (16): 386–405.

Coe, Neil, Peter Dicken, and Martin Hess. 2008. "Global Production Networks: Realizing the Potential." *Journal of Economic Geography* 8 (3): 271–295.

Dunning, John. 1998. "Location and the Multinational Enterprise: A Neglected Factor." *Journal of International Business Studies* 29 (1): 45–66.

Eriksson, Rikard and Urban Lindgren. 2009. "Localized Mobility Clusters: Impacts of Labour Market Externalties on Firm Performance." *Journal of Economic Geography* 9 (1): 33–53.

Ernst, Dieter and L. Kim. 2002. "Global Production Networks, Knowledge Diffusion and Local Capability Formation." *Research Policy* 31 (8–9), 1417–1429.

Fair Labor Association. 2012. *Independent Investigation of Apple Supplier, Foxconn: Report Highlights*, http://www.fairlabor.org; accessed September 5, 2012.

Fan, C. Cindy. 2011. "Settlement Intention and Split Households: Findings from a Survey of Migrants in Beijing's Urban Villages." *The China Review* 11 (2): 11–42.

Flamm, Kenneth. 1984. "The Volatility of Offshore Investment." *Journal of Development Economics* 16 (3): 231–248.

Gereffi, Gary. 1994. "The Organization of Buyer-Driven Global Commodity Chains: How US Retailers Shape Overseas Production Networks." In *Commodity Chains and Global Capitalism*, edited by Gary Gereffi and Miguel Korzeniewicz, 95–122. Westport, CT: Praeger.

———. 1995. "Global Production Systems and the Third World Development." In *Global Change, Regional Response: The New International Concept of Development*, edited by B. Stallings, 100–142. New York City: Cambridge University Press.

Gereffi, Gary, John Humphrey, and Timothy Sturgeon. 2005. "The Governance of Global Value Chains." *Review of International Political Economy* 12 (1): 78–104.

Giovannetti, Giorgia and Marco Sanfilippo. 2009. "Do Chinese Exports Crowd-Out African Goods? An Econometric Analysis by Country and Sector." *European Journal of Development Research* 21 (4): 506–530.

Hess, Martin and Henry W.-C. Yeung. 2006. "Whither Global Production Networks in Economic Geography? Past, Present, and Future." *Environment and Planning A* 38 (7): 1193–1204.

Hille, Kathrin. 2011. "Foxconn Looks to a Robotic Future." *FT.com*, August 1, http://www.ft.com/cms/s/2/e5d9866e-bc25-11e0-80e0-00144feabdc0.html#axzz2K41NaquD.

Klevorick, Alvin. 1996. "Reflections on the Race to the Bottom." In *Fair Trade and Harmonization, Prerequisites for Free Trade*, vol. 1, Economic Analysis, edited by Jagdish Bhagwati and Robert Hudec, 459–467. Cambridge, MA: MIT Press.

Krugman, Paul. 1991. *Geography and Trade*. Leuven, Belgium: Leuven University Press.

Li, Yu. 2004. "Education and Development: A Historical Experience of Sichuan." In *China's West Region Development: Domestic Strategies and Global Implications*, edited by Ding Lu and William Neilson, 309–332. Singapore: World Scientific Publishing Co.

Marshall, Alfred. 1920. *Principles of Economics*. London: Macmillan Co.

Martin, Ron and Peter Sunley. 1996. "Paul Krugman's Geographical Economics and Its Implications for Regional Development Theory: A Critical Assessment." *Economic Geography* 72 (3): 259–292.

McNally, Christopher. 2004. "Sichuan: Driving Capitalist Development Westward." In *China's Campaign to 'Open Up the West': National, Provincial, and Local Perspectives*, edited by David Goodman, 112–133. New York City: Cambridge University Press.

Mehmet, Ozay and Akbar Tavakoli. 2003. "Does Foreign Direct Investment Cause a Race to the Bottom?" *Journal of the Asia Pacific Economy* 8 (2): 133–156.

Pfeffer, Jeffrey. 1981. *Power in Organizations*. Boston, MA: Pitman Publishing.

Porter, Michael. 1990. *The Competitive Advantage of Nations*. New York City: Free Press.

Pun, Ngai. 2005. *Made in China: Women Factory Workers in a Globalized Workplace*. Durham, NC, and London: Duke University Press.

Rainnie, Al, Andrew Herod, and Susan McGrath-Champ. 2010. "Workers in Space." In *Handbook of Employment and Society: Working Space*, edited by Susan McGrath-Champ, Andrew Herod, and Al Rainnie, 249–272. Cheltenham, UK: Edward Elgar.

Riisgaard, Lone and Nikolaus Hammer. 2011. "Prospects for Labour in Global Value Chains: Labour Standards in the Cut Flower and Banana Industries." *British Journal of Industrial Relations* 49 (1): 168–190.

Rudra, Nita. 2008. *Globalization and the Race to the Bottom in Developing Countries: Who Really Gets Hurt?* Cambridge, UK: Cambridge University Press.

Tsui, Enid and Simon Rabinovitch. 2012. "China Pushes Minimum Wage Rises." *FT.com*, Jan. 4, http://www.ft.com/s/o/847b0990-36a2-11e1-9ca3-00144 feabdc0.html/#axzz2L3NVIR7c.

Tu, Lufang. 2012. "China's Logistics Cost Totals 8.4 Trillion Yuan in 2011." *People's Daily Online*, Feb. 16, http://english.peoledaily.com.cn/90778/7731408.html.

Wang, Zhikai. 2011. "Social Security for China's Migrant Workers." *International Labor Review* 150 (1–2): 177–187.

Wei, Yehua Dennis, Wangming Li, and Chunbin Wang. 2007. "Restructuring Industrial Districts, Scaling Up Regional Development: A Study of the Wenzhou Model, China." *Economic Geography* 83 (4): 421–444.

Wenweipo. 2012. "Huang Qifan: nuli cujin Chongqing jiagong maoyi dafazhan (Huang Qifan: Working Hard to Promote Processing Trade in Chongqing)." Aug. 16.

Williamson, Oliver. 1975. *Markets and Hierarchies: Analysis and Antitrust Implications.* New York City: The Free Press.

———. 2008. "Outsourcing: Transaction Cost Economics and Supply Chain Management." *Journal of Supply Chain Management* 44 (2): 5–16.

Wu, Fu-Xiang. 2012. "Chongqing bidian butie zuikuai hounian tuichang (Chongqing's Subsidies to Notebook PCs May End in 2014)." *United Daily News,* Aug. 20, http://paper.udn.com/udnpaper/PIA0028/221893/web/.

Wu, Jieh-min. 1997. "Strange Bedfellows: Dynamics of Government-Business Relations between Chinese Local Authorities and Taiwanese Investors." *Journal of Contemporary China* 6 (15): 319–346.

Xu, Chenggang. 2011. "The Fundamental Institutions of China's Reforms and Development." *Journal of Economic Literature* 49 (4): 1076–1151.

Yang, Daniel Y.-R. and Neil Coe. 2009. "The Governance of Global Production Networks and Regional Development: A Case Study of Taiwanese PC Production Networks." *Growth and Change* 40 (1): 30–53.

Yeates, Nicola. 2004. "Global Care Chains." *International Feminist Journal of Politics* 6 (3): 369–391.

Yeung, Henry W.-C. 1998. "Capital, State, and Space: Contesting the Borderless World." *Transactions of the Institute of British Geographers* 23 (3): 291–307.

———. 2005. "Rethinking Relational Economic Geography." *Transactions of the Institute of British Geographers* 30 (1): 37–51.